This book is dedicated with respect
to the memory of

R.D. Lawrence

Naturalist, outdoorsman, author, long time advocate for the wolf, and Friend of Haliburton Forest and it's Wolf Centre.

"We cannot smell the world of the wolf, we cannot hear it, we certainly do not see it in its proper perspective, we cannot taste it and we do not know how to come to terms with it. Instead, we blunder through it, believing that we are the masters of creation and refusing to accept that the "lower animals" are constantly in touch with the realities of the universe."

R.D. Lawrence
In Praise of Wolves

CONTENTS

By Way of Introduction   page 2

Meet the Wolf   page 5

Family Matters   page 11

Through the Ages   page 21

Chance Encounters   page 29

Discover Haliburton Forest   page 39

Wolves at the Centre   page 51

Use Your Imagination   page 61

Finding Out More   page 75

Support the Wolf   page 79

You Were Saying   page 83

Acknowledgements   page 87

By Way of Introduction

Few people have ever had the privilege of seeing wolves in the wild. Those that live in wilderness country are more likely to gain a glimpse of these elusive creatures but even then the view will be brief. Most of us who do encounter them will have a rare sighting, perhaps a wolf skirting the edge of a frozen lake or fleetingly caught in headlights while crossing a bush road. Our memory may instead be of tracks in snow or a howl floating on the night air.

True wilderness is not easy to find any more and so the wolves retreat to treed bush and tundra country where they are unlikely to be bothered by human beings. The Canadian Shield, that great arc of rock and glacial gouges that is centred in northern Ontario and Quebec and has the highlands of central Ontario as its base defines for us the possibility of seeing wolves. South of the shield are automobiles and people, both species shunned by wolves. Enter that rocky upland country and move through wolf country.

Most of us rely on others for news and knowledge of wolves. Scientists, naturalists and dedicated observers fill in the gaps in our casual acquaintance with wolves. Wildlife photographers, generously endowed with patience and a hardy approach to life, provide us with images of this creature that has captivated much of mankind since the beginning of recorded time.

There are no absolutes in the wolf story. Those who speak most wisely of the animal do so in a series of cautions. They avoid generalizations and tell us that all wolves are different and that this most adaptable of predators never ceases to demonstrate different behaviors to challenge long held notions.

Although most of our knowledge of wolves comes second hand via written accounts and carefully obtained photography, there is one way to get to know them better. In a few large scale sanctuaries, preserves and centres, on the North American continent, the wolf pack may be seen by the patient observer in close approximation to its life in the wild. One of finest of these is found in the Wolf Centre at Haliburton Forest and Wildlife Reserve.

Here is held the only pack of unsocialized wolves in North America. This simply means that while these magnificent animals are captive in a very large enclosure, they do not associate with humans and are presented for observation as near as possible to life in the wild.

There are hundreds of books on wolves. This modest account has very specific aims. Generally accepted knowledge garnered by experts in the field over the past sixty years has been synthesized for the busy but interested reader. The wolf is then illustrated through human attitudes over time. Some varied encounters with wolves are presented. Next follows the story of the recovery from years of neglect and abuse of Haliburton Forest, the home to the Wolf Centre which has become a mecca for animal lovers across the continent. The wolf in the imagination provides an opportunity to rediscover how wolves have been presented to us and perhaps recall our own meeting with wolves via a variety of media. The reader is lastly offered some ways to find out more perhaps even support the continued survival of wolves and reflect on what others have said of the animal.

At the end of this brief encounter with wolves, one might paraphrase naturalist Henry Beston who said that wolves are not our brothers; they are not our subordinates, either. They are another nation, caught up just like us in the complex web of time and life.

Meet the Wolf

The wolf has survived over time because he is versatile, adaptable and quite possibly the most intelligent being in the forest. A highly social animal among his fellows, an affectionate father and good provider, he is extremely cautious towards others than his own kind. He operates in groups and has mastered that hallmark of advanced societies, a well developed division of labour in everyday communal living. A highly efficient system of various means of communication helps him to achieve his maximum potential. Take a look now at the one who may be the most efficient predator on earth next to man.

Linnaeus, Carl von Linne the Swedish scientist who lived from 1707 to 1778, coined for the wolf the scientific name *Canis Lupus*. Canis is from the Latin for dog. Practically all dogs are evolved from wolves. Wolves are simply large wild dogs although their behaviour is similar to the domestic variety. There are three main species in the world, and scientists differ about the number of sub species. Those found in North America are the gray wolf, the most common, the red wolf (found in the southern United States) and the Algonquin wolf domiciled around the great Ontario park of the same name. Wolves in this book generally refer to the gray or timber wolf. The wolf originated in North America and crossed the former land bridge to Asia. Later the animal returned to this continent, thus the division into the "red", the original and "grey" the returning wolf.

The so called "brush wolf" is not a wolf at all, but a coyote. These animals are much smaller than a wolf and have a pointed smaller face. They generally have thick tawny fur and operate as scavengers, skirting human communities looking for food and even killing domestic animals. They do not howl but have a cry somewhat like a dog's yelp.

Dogs and wolves are social animals and are similar in many ways. Dogs need and willingly obey people whereas, while social among their own kind, wolves shun human contact. Dogs act in a spontaneous fashion, rushing into situations, whereas the wolf is ever cautious, always reconnoitering the scene around him. Dogs suffer from emotional disorders due to selective breeding for the convenience of their owners. Wolves have an aristocratic bearing. The fineness and delicacy they possess has been lost to dogs after centuries of breeding for specific purposes. The dog's brain is 31% smaller than that of the wolf. A simple example of intelligence or rational thinking observed in the wolf is that one animal in captivity could open a door after having watched humans over time performing the same operation. At the Haliburton Forest Wolf Centre, an Alpha male examined the plastic-coated clothesline used to tether a beaver carcass at feeding time. He traced the line and eventually cut it and released the meat. The future use of steel cable put an end to his original idea.

In turning to the physical characteristics of wolves, various authorities have debated on the most outstanding feature. This generally comes down to their most expressive faces. The eyes are most commonly noticed. As naturalist R.D. Lawrence put it, 'No one can deceive the eye of a wolf: variously intense, scouting, probing, appraising one for fear, aggression and honesty'.

Wolves make the most of their visible capabilities as a mainly forest dweller because their lives depend on it. Males are generally a fifth larger than females. Due to their larger size they usually initiate the attack on a hunt. Bigger on the average than dogs, male wolves weigh between 70 and 110 pounds, have an average length from nose to tail tip of between 5 to 6.5 feet and their height is 26-32 inches. Their track varies from 4 ½ inches long to 2 ½ inches wide; only a few breeds of dogs leave tracks approaching this size. This unique package of adult size is reached within one to two years and from then on they may gain weight but size remains the same. Attaining full size early is vital for the animal to travel in search of food.

In exceptional cases wolves can outweigh humans, although in this historic depiction, the animal may be a dog-wolf hyubrid

Long legs aid in running and seeing far across the land. The great brushy tail helps hold balance and making fast turns on the run. Those large feet that spread on impact are great natural snowshoes. A loping run on toes allows fast stops and turns. With knees in and paws out, the hind legs have a path to follow which results in a single track. Wolves may travel for hours at about five miles an hour but can put on 25 to 35 m.p.h. for short periods when chasing prey. These animals have great stamina. Try traveling an average of 30 miles a day. Wolves do this all the time when in search of food. Then they are on the move 8-10 hours out of every 24, usually out on the hunt in the crepuscular or twilight hours.

The variety of wolf pelage- fur or hair- is greater than any other mammal. The most common colour is a tan or grizzled gray but specimens may run through from glistening white to coal black. One Arctic wolf was observed as being black with white patches. The coat is shed in summer so that a new one is grown ready for cold months. There are two layers. Coarse guard hairs repel moisture while the soft under fur acts as insulation. Faced with a bitter cold wind at low temperatures, the well equipped wolf places his face between his rear legs and covers it with his bushy all purpose tail. In common with the wolverine, wolf fur does not ice up when met by warm breath. Another major cold weather advantage enjoyed by the wolf is the ability to reduce the flow of blood near the skin to conserve heat. In this manner the temperature of paw pads may be maintained above the freezing point.

We possess 32 teeth but adult wolves have ten more. By comparison a mountain lion has only 30. Long legs get the wolf to its prey and with luck its teeth finish the job. The four great pointed canines work to seize flesh, pierce it and hold on to a struggling quarry. The pre-molars tear and the molars crush. Incisors cut and tear meat from the bone. Some molars are carnarsial teeth that sheer like scissors while other flat molars grind or pulverize bone. Put all this power together and the molars and jaw can attack bone at 1500 lbs per square inch of pressure, which is really great for getting at the marrow. Under this form of attack, a moose femur can be severed in six or eight bites. By contrast a German shepherd has only half that awesome power while humans can only manage 300 lbs per square inch.

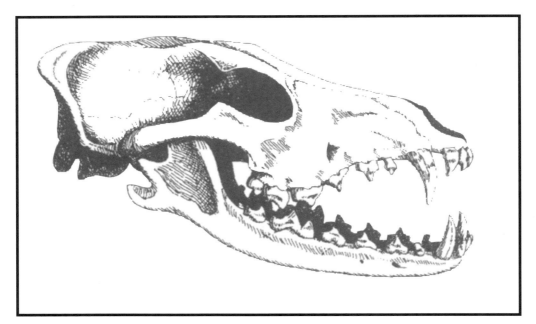

The Functions of the Wolf Skull:

The wolf skull has 4 canines which hold on to the prey, incisors in the front shear meat and sinew, premolars and molars grind the meat in preperation for digestion. The large bone ridge at the back of the skull allows strong muscles to transfer power to the jaw.

A wolf in the wild uses all of its senses to the utmost. The sense of smell is the most acute. A wolf's nose has a surface area 14 times that of man which means his ability to smell is 100 times that of his two legged fellow mammal. This sense sure helps in detecting aggression or alarm in others. Hearing is four times the capacity of humans and with the right conditions sounds up to six miles away may be picked up in the forest. Vision is the least developed sense but the animal makes up for this by extreme sensitivity to movement. Night vision is stronger than that in humans and peripheral vision is excellent. Wolves have detected Arctic hare, a gray movement on gray green tundra hummocks, and run it down over a quarter of a mile stretch using a combination of their vision and ability to detect distant motion.

There is an outline of the physical aspects of wolves found in the Haliburton Forest Wolf Centre. Now move on to meet the family.

Wolf Centre pack member Manitou smelling for a hidden cache of food.

This early British colour postcard printed in Saxony presents the wolf well. A note that when hungry the wolf will enter houses and attack inmates may be taken with a grain of salt.

Family Matters

Wolves are highly social animals that generally live in extended families we call packs. In the wild in the Haliburton area these average 3 to 7 members and they may require free range in up to 60,000 acres to sustain life. The strongest personality trait among wolves is the capacity to associate in an amicable way with others of their own kind and as a result wolves are always looking to enroll in a pack. Within this customary group are found male and female leaders, the Alpha pair. Subordinate to them will be non breeding Beta males and females, yearlings pups and on occasions outcasts living on the fringe of the main body. Wolves are generally friendly among each other and depend for survival on cooperation. They are patient and highly adaptable but there is a distinct 'pecking' order as they communicate and interact.

Wolves are spread out over territories that vary in size and range according to the availability of game, water and shelter. Their lifestyle dictates that they live in low densities. The environment in which they live determines their potential quality of life. Each animal develops a distinct personality and members even appear to the observer to 'smile' as they relax within the family unit. They may turn out to be aggressive or shy. Adults will even play like pups on occasion.

The Wolf Centre pack in the midst of a family greeting.

In common with humans they will sometimes go off for a while just to be alone. Running through all the associations the pack members have as a group is a definite submission to the Alpha male or female. This is indicated in body position, movement, in licking the face of the leader or rubbing up against him. The Alpha's actions ensure he receives the attention and deference of the pack and his position maintains the cohesion of the group. Often a female Alpha will lead a pack. Her decision regarding den location largely influences hunting grounds for several weeks.

The pack renews itself when the Alpha pair mate. In the wild too many mouths to feed in a social group can mean disaster in times of prey scarcity. A form of birth control is thus practiced where normally only the leaders of the pack produce offspring. The Alpha female will actually intimidate and stress other females to discourage any sexual activity. Her own conception will occur in late February and gestation takes 63 days. The den selected might be one used before, in a rock overhang, hollow log or cleft in rocks. Most commonly the location of choice is a short tunnel dug by the pregnant Alpha, with a dogleg turn leading to a hollowed out area where the pups weighing less than one pound are born.

One week old wolf pups in a den at the Haliburton Forest Wolf Centre in 2002.

Noted observer Lois Crysler says wolves are crazy about pups. The pack gathers around the den pacing up and down excitedly at the time of the birth. Staff at the Haliburton Forest Wolf Centre note the change in pack dynamics even though the birth den is out of sight. The litter may range from 1 to 12 pups, which are born deaf and blind, being able to hear in a few days and opening their eyes from eleven to fifteen days. By five weeks they are weaned and venturing out of the den, being moved by the adults to another secluded location. Mortality may be up to 90% in the first year due to scarce food, disease or the attacks of predators who evade the mother or other wolves acting as baby sitters. By 12 weeks, those who survive will travel with the pack but not join the hunt until 7-8 months. All the time they are schooled, disciplined and played with in strong bursts of affection. They receive food still warm from the kill, regurgitated from the bearer's mouth. Gradually the youngsters develop individual personalities, and in typical wolf fashion, a few begin to dominate other siblings.

Some yearlings stay with the pack while others leave to find new groups or start their own. There are lone wolves who have been ousted from a pack or find themselves alone due to perils encountered in the wild. Such unfortunates are rare and lacking the hunting efficiency of pack life, they usually try to join other wolves. Some wolves could live from nine to seventeen years but the average lifespan is more normally five years in the wild. The stresses of survival, getting food, avoiding man and a host of diseases plus the menace of other predators all tend to shorten the lives of wolves. During their life they are always active, learning new skills and in packs always endeavoring to pass this knowledge on to younger animals.

Wolves maintain the pack with a variety of ways of communication with each other. All of these are paramount in hunting. By observing these postural and vocal signals, humans find wolves easy to study when attempting to figure out the current status within the social group. The tail is a good indicator. The higher a wolf's spirits, the higher goes its tail. Eye contact or lack of it between animals reveals much. A direct stare is interpreted as a threat while an averted gaze and turning away from the other member denotes submission. A social invitation is offered with ears forward, a play 'grin' on the face and a wagging tail. Moves by a leader to promote dominance may include growling, a direct gaze, bared teeth and nipping or seizing the muzzle of a subordinate. A tail down, flattened fur and head lowered are some of the ways a lesser animal acknowledges his or her leader. There are many gestures and expressions not fully understood by wolf watchers and these examples are just a few of the postural signals in the wolf arsenal. Wolves know what they all signify and as a result have the assurance which allows them generally to live in peace with their fellows.

A howl is a happy occasion as wolves love to communicate. They do verbalize with whines, whimpers, yelps, and short barks but the various individual and group howls are like parts in a community singsong. Wolves have many postures but the favorite one seems to be driving their hips forward and pointing the muzzle to the sky. The wild, beautiful and eerie cries are what one researcher called 'the jubilation of wolves'. The animals enjoy howling but often have distinct aims for these highly vocal expressions. They may be used in bringing the pack together, signalling the start of a hunt, attracting a mate or to signal a warning. Giving such voice may indicate anxiety but each effort, which some Indians thought was the spirits talking, does have varying effects on humans who may hear the sound up to ten miles away. The writer observed a group of visitors to the Haliburton Wolf Centre in mid morning without an animal in sight. A brief howl by a staff member quickly brought most of the pack close; curious to know what was happening. One of life's great experiences is to take part in an organized wolf howl and receive the call back across the wild. Such events take place at both Haliburton Forest and Algonquin Park.

The wolf's howl has always captured man's imagination and curiosity.

Wolves have a third means of getting in touch in addition to body language and vocal signals. They make contact using the sense of smell. The customary pack territory is marked at well established intervals on trees, rocks and other familiar landmarks that outline the perimeter and other parts of their land. These markings or scent posts work as a map as well as a clear warning to strangers and can offer a variety of useful bits of information. Such 'notice boards' are visited regularly to keep the scent current and thus the raised leg urination had a significant purpose other than waste relief. Scientists tell us that wolves actually develop an olfactory memory as one tool to help figure out the world in which they live.

Wolves communicate in well determined ways but obtain food as opportunists. Their digestion and way of life leads them to adapt to a feast and famine existence. Such a way of life can mean days without food and then a gorging binge on large amounts of meat obtained by their own endeavors. One fifth of individual body weight may be eaten and processed in a few hours. In tough times mice, other rodents and even berries may represent the next meal. These predators that kill for a living prefer to feast on ungulates (hoofed animals) such as deer, moose, elk and caribou. Beaver is also a welcome meal. Such is their diet at the Haliburton Forest Wolf Centre.

While on the move wolves are also hunting. They take in every visual and sound sensation and check all possible prey locations in their well known territory using smell, tracks and chance encounters of food sources. Sometimes they seem to test a deer herd, making themselves known so that old weak or nervous animals may be detected. The hunter may start a chase looking for the same type of quarry. There is no set pack plan of attack; instinctively each member knows what to do in concert with his fellows. Decoy and ambush are often used tactics. The wily hunter may try to unsettle prey in winter by driving it out on the ice or on stone filled dry creek beds in warm weather months. When a prime target is chosen there is no assurance that the hunt will be successful, in fact on the average only one in ten such ventures results in food for the pack. Careful to keep away from rearing hoofs or well positioned antlers, the larger male wolves go first and will slash at flanks, tear at hind quarters or hang on to vulnerable heads. Overall stamina and endurance aids the persistent hunter to keep up with an often faster prey. The wolves are realists and will discontinue the hunt when the odds are against them.

Successful hunting trips ensure pack survival. The Alpha pair feeds first, slashing open the soft under belly, seeking the intestines and other nutrient loaded warm organs first. All family members then share in the feast. They will bring meat to pups and less able pack members. Practically all the downed animal will be consumed and when there is a surfeit of meat the balance will be cached for another day. A kill is only abandoned when the wolves are disturbed. Too often wolves have to compete with humans for the same prey and they are made scapegoats by disgruntled, unsuccessful human hunters.

In all their interactions in the wild, wolves seem to have a special relationship with ravens which is both practical and casual. Some observers believe the birds actually lead the wolves to dead animals or potential prey. By performing initial butchery, the wolves open up a carcass which may eventually prove a feast for the patient birds. The wily ravens are not above the risky business of stealing such food and have even been known to tweak a close by tail to distract the wolf from his dinner. When hunger is satisfied, a game of tag sometimes occurs where the birds venture in close and just as quickly fly out of the reach of the powerful animal they are teasing. For the ravens, this activity is truly living on the edge as annoyed wolves on occasion react faster than the bird can take off. Even a bird may be consumed.

Ravens have been known to follow wolf packs in search of an easy meal of leftovers.

So the wolf pack attends to family matters and makes its way in the wild, avoiding man who ironically after centuries of persecution is now working in many areas to protect wolves. As ever the pack is versatile, an adaptable unit seeking to live in the forest and live as wolves have always done. Fortunate are well intentioned people who come across wolf tracks and signs and hear their howls, knowing that a vital creature of the wilderness is still thriving. They may see evidence of this in the Haliburton Forest.

In this picture of wolf tracks near Marsh Lake one can clearly observe the tracks of two animals "overlapping" in the foreground.

An 18th century Riedinger engraving of a seven member wolf pack.

Wolf Centre pack member Citka in a full on yawn,
expressing a sense of relaxation.

In this 1930's National Geographic print the Peary or Barren Ground Caribou is in a good position. He has his back to the rocks and is facing the attacking arctic wolves from above.

Through the Ages

Throughout all but earliest history the wolf has been the unfortunate victim of a bad press. From earliest times the animal has been reviled, libeled and slandered, been arbitrarily and generally unfairly given the attributes of greed, viciousness and cunning. Practically any calamity or evil unleashed in the world has at some time or another been associated with the wolf. Take a look at some of the human thinking about and treatment of wolves over the centuries. Some of these ideas will be introduced again later.

The first cave drawings depicting wolves did not appear until about 20,000 years ago in Southern Europe. In the Miocene era, twenty million years ago, distinct dog and cat families of carnivores were established and the ancestors of the wolf had emerged. Forerunners of the present wolf were no doubt cutting out weak, lame or young camels in what is present day Oklahoma. The ancestors of present day wolves were the creatures from whom coyotes, jackals and foxes evolved.

Positive accounts and legends about wolves from earliest times were gradually forgotten as a climate of fear and superstition grew up with the wolf as the main target. As early as 300 B.C. the Celts were breeding wolfhounds to kill wolves and there is a steady progression of extermination right up to the twentieth century. January in England was Wolfmonat or Wolf Month. This was a time when wolf slaughter was specially focused. Wolves depicted in fables by writers like Aesop, la Fontaine, and Krilov generally portray the hapless animal as cunning, greedy and sinister. At best the wolf is portrayed in a one-dimensional way illustrating one virtue or more often, a cardinal sin.

Once man began to live as a farmer and herder, the wolf came to be portrayed as an out-right killer. As settlement moved forward, wolves were forced out and retreated. In the Bible, John 10:12 has the disciple relating that Jesus described himself as a shepherd protecting the sheep from the wolf. This image of the wolf as an enemy of man was perpetuated by the church over the centuries and fostered hatred of the animal. One old saying had it that the wolf did such bad things during the week that prevented him from going to church on Sunday.

Shakespeare equated the wolf with hunger, 'the universal wolf'. So it is no great stretch to such modern sayings that a family should stock up to 'keep the wolf from the door'.

In the middle ages the wolf was the repository of all human fears, the scapegoat to human ignorance. During the Inquisition when men were burned at the stake, one of many charges against them was that the unfortunates were werewolves. Once more the wolf was equated with evil. The Thirty Years War which ravaged much of Europe from 1618 to 1648 saw vast amounts of humans left dead on far flung battlefields. Wolves took advantage of this unexpected bounty and were seen by peasants as they ate the war dead. It was not a big stretch from this to have stories abound that wolves killed men. Beliefs of those dark times were all against the wolf. If a horse stepped in a wolf print it was felt that it would be crippled. The word loophole comes from Loup or wolf hole, a place where travelers might watch for wolves.

Many cultures felt there were half man, half animal creatures which were known as werewolves. These supernatural beings were current in popular belief from the earliest to modern times. The movies and lurid potboiler novels have made great profits from this convenient peg on which to hang vague fears and superstition. In the voodoo of Haiti, the Loup Garou is the werewolf representative of Satan. There are many current books about werewolves, testifying to the still popular morbid curiosity regarding these beasts of the imagination. Lycanthropy is the psychiatric disorder in which a patient believes he or she is a wolf. Hallucinations of this type could be caused by drugs like belladonna, hemlock, aconite and opium.

Hollywood loves the "Wolf Man" character. In this 1940's movie version the character is represented by actor Lou Cheney.

False beliefs about danger from wolves have picked up on isolated incidents and magnified them through the telling. Incidents involving rabies have added fuel to fears of wolf attacks. Over the past century people like Captain Lewis of the Lewis and Clark expedition have found wolves to be gentle. Wolves have long been known to be one of the shyest animals in the wilderness and they will go to great lengths to avoid man. Famed arctic explorer Vihjalmer Stefansson tracked every report of a wolf killing a human between 1923 and 1936 in the far north and could not substantiate one. Sault Ste. Marie *Daily Star* editor James Curran wrote extensively about Algoma wolves and never had to pay out on his $100 reward for evidence of a wolf attack on a human being. Despite these efforts, there have been cases of the death of humans from wolf attacks in North America.

Wolves have been hunted since recorded time. Justification for such killing was the protection of livestock, eradicating wolves so that hunters would have sufficient supply of game to gather food in winter and the sport of trophy hunting. Pioneers felt wolves were just another part of the wilderness they had to tame in order to settle the land. Teddy Roosevelt called it 'the beast of waste and desolation' and often set out on huge hunts in both North America and Russia. He ignored the fact that wolves were often blamed for damage and livestock loss caused by over enthusiastic hunters.

This painting in the Louvre by Desportes (1661-1743) has dogs used in a wolf hunt.

Wolf bounties in North America are recorded widespread between 1630 through until the 1920s. Huge numbers were killed in Alaska and mainland states from 1820 to 1920. In 1907 the US Department of Agriculture put out a helpful pamphlet, 'A Guide to Finding and Killing Wolves'. Several Canadian provinces offered bounties for killing wolves and some counties in Ontario continued the practice long after the province had ceased to promote wolf killing. Alaska came back with a big kill in 1993 and the aerial hunt where a whole pack could be destroyed by one sharp shooter was excused by declaring that food hunters could stock their larders with big game supposedly threatened by wolves.

In his time (the late 1900's), this successful hunter would have been applauded rather than condemned for taking so many wolves and coyotes.

Francis Parkman, the great American historian, wrote 'The wolves that howled at evening about the traveler's campfire, have succumbed to arsenic and hushed their savage music'. So for years wolves have been destroyed by various cruel means, largely because it was felt they were a threat to human settlement. In Ontario the wolf bounty was not repealed until 1972. Haliburton County kept the bounty until 1989 when it was as high as $100 per tail. We know today that such payments for killing wolves failed as an effective means of predator control. Ernest Thompson Seton wrote as early as 1898 of wolves that 'they seem to possess charmed lives, and defied all manner of devices to kill them'.

So where are wolves found today? Popular estimates made in 1994 set the number of wolves in Canada at 55,000, Alaska 7,000 and the rest of the mainland U.S.A. at less than 2,000. Wolves have been exterminated in the British Isles and much of Europe. There are some in the wild in northern Spain and the Italian Apennines. The Middle East and North Africa have very reduced populations and the wolf population in China is not known. An estimated 50,000 wolves roam Russia. There is a small captive population in Mexico and wolves are found now in nine mainland U.S. states. The last wolves were killed in Yellowstone National Park in 1926 and since 1995 when they were reintroduced from Canada after much public debate, the packs there are surviving. In Idaho the Nez Pierce Indians were so anxious to support wolf recovery that the tribe took over the support program for the animals their people had admired through the centuries. The fact that wolves were declared to be an endangered species in the United States in 1973 was a call to arms for all people interested in their welfare. So the wolf is coming back in North America. However, it can only do so where there is still enough wilderness range to ensure survival is a viable proposition. Where the range of these predators does overlap livestock grazing land, several private and government groups have programs to compensate farmers and ranchers who do not maintain a secure environment for stock loss.

Wolves are still trapped but, fortunately for the animal, the fur is not in high demand, although the lighter colours are used in coat and parka trim. Alaska allows wolf hunting and in Canada, the country with perhaps the largest wolf population, there are still very liberal hunting laws. A small game license gives a hunter the right to hunt wolves year round in Ontario. Much work is done now on wolf research and scientists are gaining a much clearer picture of the lives of these elusive animals. The general public is now oriented towards conservation and a very important means of bringing wolves to prominence has been proper display. Zoos endeavor to educate visitors but by virtue of their size and organization often do little more than entertain a curious public. Wolves on display in such places are not allowed to develop using their high intelligence, receive mental stimulation or room for their pack hierarchy to function. Fortunately there are now places like the Haliburton Forest Wolf Centre and other sanctuaries and parks that have enough free range to offer a glimpse of the wolf in a wilderness setting and garner support for the future of this still little known animal.

This prairie print refers to wolves as "cowardly" and "pirates of the plains", ignoring the fact that bringing down a strong, old buffalo bull would be a daunting task for any wolf pack.

Many old prints like this illustrate the perceived cunning of the wolf.
As the shepherd and his guard dog sleep, the wolf approaches the herd in disguise.

Depicted in this print is the middle-aged custom of punishing dirty language by placing a wolf mask on the offender before putting him on public display.

Wolfers were men who killed wolves for bounty in the U.S.A.
This is the historic view into a "wolfer's roost".

Chance Encounters

Alaska holds the dubious distinction of allowing aerial hunting up to less than thirty years ago and the practice has still not been stamped out. Barry Lopez tells the story of one such 'sporting' endeavor in the winter of 1976. A gunner and his buddy were flying low in the Alaska Range. Suddenly they came across a pack of ten gray wolves traveling in open country along a high ridge. Calmly and methodically the rifleman killed nine of the animals but the animal who was now the last of his pack made for an abrupt vertical drop of perhaps 300 feet. The plane swooped in but the man who had decimated a wolf family was curious as to what would happen next. Endowed with the courage and determination that accompany his species, the wolf launched off the high point and fell that sheer drop into a snow bank. In a spray of powder snow he disappeared into the trees and the aircraft veered off back to its base, where the shooter could boast about his success. As leader, the survivor was likely the Alpha male of his pack. There is no known end to this encounter. Maybe he was able to get another group of wolves together. Maybe he just became a lone wolf and died of starvation.

The great American poet, naturalist and essayist Henry David Thoreau once offered some advice for amateur wilderness watchers. He said that over time you learn that if you sit down in the woods and wait, something happens. Each encounter with wolves, however fleeting, adds to our store of knowledge about these animals. Presented here are a few reminiscences about wolves by people who did spend time in the woods and waited for something to happen. A few originate in the U.S.A. but the majority took place in Ontario and the Haliburton Forest area.

Famed geographer and explorer David Thompson actually mapped the northern part of Haliburton County. He noted in his diary having observed two wolves bringing down a deer. In this instance they bit at the animal's hind legs until sinews were cut. No slouch himself in the attribute of persistence, Thompson was impressed by the way the two predators shared the hunting role and how they kept on until the job was done.

Images of wolf hunts like this Wapiti or Elk chased on the ice and falling prey to the pack were common in early literature.

The idea persists among people who have never seen a wolf and even those familiar with the animal that they are somehow bigger than real life evidence confirms. One Minnesota trapper looked over a trapped wolf and declared it to be upwards of ninety pounds. When it was weighed and found to be sixty seven pounds, the woodsman was quite indignant. He felt his catch should have been much bigger and opined that it must have been sick.

Stories abound that wolves attack humans. Few may be verified. A man named S.C. Tumbo of Springfield, Missouri was reliably said to have been treed by a pack of wolves. When they were dispersed later he came down and puzzled over the incident. A friend pointed to the likely cause. His coat was drenched with blood from an earlier hunting adventure.

Trapper Len Holmes of the Wilberforce area east of Haliburton had a once typical bushman's hatred of wolves. He felt that they took deer he might hunt and yet had to admit that there were still plenty of deer to go round. His encounter with wolves when a teenager, where he could 'hear their teeth snappin' in the bush, left a lasting bad impression. They interfered with his traps yet he was glad to receive bounty money when he trapped one. By contrast, another Haliburton trapper, Theo Peacock of the Gooderham area was perhaps more sensitized to wolves' place in nature. He wrote a poem called Mother Love about a wolf and her long term grief after her pups were killed by hunters.

Examples of wolf play in the wild are well documented. One Alaska sighting was more in the way of a prank taken at the expense of another species. Three wolves came over the side of a hill in the Brooks Range and saw amid the sunlight on a glacial pond ahead of them a bunch of pintail ducks. They flattened out and crawled slowly toward the unsuspecting ducks. On some unnoted signal, they sprang into the air as one and bounded toward the ducks. The startled quarry exploded into the air in a blaze of colour and fast flapping wings and the wolves splashed around in the once again still water and took a drink. The brief event was just a spur of the moment game.

Pack members Cedar and Citka in a playful tussel.

In a similar vein one researcher had the good fortune to see a young male gray wolf sneak up on an eagle perched on an ice ledge. He then made a movement with his feet and stood and grinned as the eagle saw his unwanted companion and burst into the air. The yearling took a break and then spied some ducks in a pond. He made an impressive rush to the water's edge and then gave a short bark of satisfaction as the ducks paddled off in a hurry.

Ralph Bice was out with a fishing party in Algonquin Park when he had an unusual battle with a wolf. He saw a deer swimming toward shore and noted a wolf behind it. Acting quickly he positioned his canoe between the wolf and the shore and heard the deer scramble up onto the bank. The wolf turned in the water and came for the canoe. Bice struck it a glancing blow with his axe, whereupon the wolf sank its teeth in the handle. Then Bice tried to push the wolf's head under the water but he came right back up and lunged at the canoe again. This time a blow from the axe right between the eyes finished off his quarry. Back at the camp fire when his encounter was scoffed at, Bice showed the teeth marks in the wood.

Accounts of frontier perils were always popular.
Here a brave pioneer woman drives off a bold wolf.

Farley Mowat believes wolves have a highly developed sense of humor and play jokes on each other. Nova Scotia wolf watcher Jenny Ryon of the Canadian Centre for Wolf Research says interactions between pack members are rather charming. The wolves love to play with each other, chasing, nipping, jumping and wrestling together, each with a favorite game. Their play is like the hit song of summer. When they get tired of it they add a new element.

One researcher had the good fortune to see a wolf family lesson in action. An Alpha female left her pups in a secluded area and loped off on some unknown errand. Then as if something had occurred to her, she stopped on the trail and lay flat in the path. Shortly after a pup, ignoring her wishes to stay with his fellows in the litter, trotted along the path. His mother gave a low bark and stared at the errant pup. At this sign he acted almost as if he had not been thinking about this forbidden excursion and slowly meandered back to the other members of the litter. When she returned that evening, all the pups were where they were supposed to be.

Sometimes visitors to Algonquin Park will ignore well posted admonitions against feeding the wildlife. Minor incidents with wolves have occurred through such thoughtless interaction with the animals. In 1998 a wolf bit a child that had come between it and some left behind food. Park officials became quite disturbed and considered some form of punitive action. This idea was dropped when a biologist observed that the wolf's action was in line with a disciplinary or annoyance reaction that would commonly be taken by an adult wolf toward a lower ranked pack member which stepped out of line.

One story gives a glimpse of a sibling encounter north of Haliburton. A wolf traveling through the forest howls and there is no answer. His sister is a mile away and trots through the trees in the direction of the call. The siblings soon meet and walk, tails erect, slowly towards each other. They make little squeaking noises, walk around in a circle and then rub and push noses into each other's neck. They move around a bit and then stand quietly resting ones head on the other's back. Then they are gone, running in single file, wagging their tails in short brisk motions.

Peter, an Algonquin Park ranger, noted a deer carcass on a frozen lake an[d] saw two wolves close by. As they sighted him, one took off into the bush bu[t] the other edged along the ice to the deer. The ranger dropped to the ice, go[t] off a shot and hit the wolf but not seriously. The wolf, finding little purchas[e] for his paws on the ice, lay on his side and rolled away toward shore. Pete[r] killed the wolf with his next shot. He skinned the animal and took the fur t[o] North Bay to claim the bounty. He thought it might be newsworthy but ther[e] had been lots of encounters that winter. Later Peter read with disgust th[e] account in the *North Bay Nugget*. The caption ran 'Another Wolf Story'.

Haliburton treasures stories of encounters with wolves in past years. Surveyor[s] came across the animals in pioneering days. Some were treed by these inquisitiv[e] animals and most of these lonely travelers maintained a healthy respect fo[r] the predators that usually outnumbered them. One early Maple Lake farme[r] out ran a wolf pack while traveling bush trails on horse back. No doubt the wolves were more interested in the horse than his rider. Lumbermen would have fun at the expense of newcomers to the bush by telling them that the rattle of a 40 pounds boom chain would scare wolves away. For a while the greenhorns would drag these chains with them until fatigue made them abandon the unnecessary loads.

This 1906 painting, supposedly based on a real event, depicts a musician trapped by wolves, keeping them distracted while he retreats to safety.

Aldo Leopold was one of the first researchers and naturalists to study wolves. In his book, *"In Thinking like a Mountain"*, he describes an incident where a party of hunters shot an old wolf. He arrived at the body as the 'fierce green fire' was dying in her eyes. That was when he realized there was something only the wolf and the mountain knew. They respected what we might call the balance of nature. Too many deer could denude the slopes of foliage. The wolf kept life in the woods in harmony.

Researcher Jamie Dutcher spent enough time with one pack that she gained deep insights into their lives. She wrote later, 'One cannot observe a wolf pack without seeing a reflection of ourselves. They seem to possess so many qualities that we admire... but despite all that has been done to them at the hands of man... wolves forgive.'

This timber wolf with pups was pictured in a mid west zoo in 1906.

Large parties went wolf hunting as shown in this vintage Russian card.

A 1950's postcard depicting the result of an arial hunt near Ely, Minnisota.

This 1931 French hunter is dressed up for his trophy picture.

The Whipsnade Zoo in England was one of the first to make animal enclosures more natural and realistic but the location still inhibits the captive wolves.

Discover Haliburton Forest

This brief introduction to the wolf now pauses to record the survival and present state of Haliburton Forest, home to one of the most innovative wolf sanctuaries in North America.

The land covered by Haliburton Forest today was formed on the southern end of the Precambrian or Canadian Shield country. Glaciers ground, scourged and scraped the rocks of the Haliburton Highlands. The retreating ice masses left behind rolling hills, hundreds of lakes and rivers with an abundance of clean water, and a base of sand, gravel, rock and boulders.

Undisturbed by man, huge forest tracts grew up in the area and the most predominant tree was the majestic white pine which covered 15% to 20% of the land. Once the rich 'front' lands, the areas just north of Lakes Ontario and Erie, were taken up for settlement, the government of the day looked to Central Ontario as a prospective home for further newcomers. Colonization roads were built to access the new territory. The British government gave a monopoly in land sales of over half a million acres in the Highlands to The Canada Land and Immigration Company which was to cover its expenses by selling land to new settlers and set them up on lots laid out in both villages and rural areas.

Surveyors were first on the ground and the most positive parts of their field reports were used to promote the land on offer and in 1864 the first settlers came to the Minden area and then Haliburton. It was not long before the newcomers found that much of the land was not suited for agriculture except for some areas in the south of the present county. The Land Company had its problems because the Crown undercut its commercial efforts by giving out free land grants and much of its holdings were leased for lumbering. No wonder that this early enterprise struggled along and was defunct by 1940. A railway reached Haliburton village in 1878 and provided service for one hundred years. In the end it ceased operation as there was not enough passenger and freight traffic to make it a viable proposition. So the Haliburton Highlands was populated very slowly and even today is only home to about 15,000 permanent residents, even though it is the third largest county in Ontario.

From the eighteen sixties through until about 1930, the vast timber tracts of the Ottawa Valley and Central Ontario fed white pine to hungry mills of North America. Much of the choice trees went for masts for the Royal Navy. The rest was used for all manner of wood products. Men commonly called timber barons made huge fortunes from this trade. The most well known of these entrepreneurs was J.R. Booth of Ottawa and the most prominent in the Highlands was Mossom Boyd of Bobcaygeon. Softwood logs were floated via the lakes and rivers south in annual 100-mile river 'drives' to Bobcaygeon. The hardwoods were spared until roads and the railway were established to enable the harvest of this valuable resource. Throughout it all there was no thought to replanting and renewing the forest. To many it seemed that the timber harvest would go on forever.

In 1869 lumbering began around Little Redstone Lake just south of the present Haliburton Forest Base Camp. Gradually these operations expanded and by 1884 the prime white pine had been largely logged out. As the Land Company ever teetered on the brink of bankruptcy, it sought to access more timber in the far northern reaches of its acreage. To do this one whole township was traded to the Crown in return for having a road built through West Guilford to Redstone Lake. With this new means of access, the Company sold some lots for the rapidly developing tourism industry and a large portion to Hay & Company of Woodstock which processed yellow birch for veneer. Some of this was actually used during World War II in the construction of the mainly birch plywood, Mosquito fighter-bomber.

The mill that commenced operation in 1944 continued cutting choice timber from the Forest until 1970, sawing over 150 million board feet in that period. It had a Base Camp with all buildings necessary for the mill, including a garage, stables, barns and other structures. The present restaurant gets its name from the original cookhouse and the office building now used by Haliburton Forest was the former mill office. The buildings comprising the central base of operations covered about 110 acres. Visitors today may acquaint themselves with the lumber industry via the display of artifacts from the period now displayed in what was once the former planer building across from the office.

The logging museum reminds us of past times in the Forest.

The mill owner was primarily concerned with profits, not in preserving or renewing the forest. There was little planning and timber was mainly high-graded, meaning that only the best trees were cut and the rest of the stand were ignored. Then Hay & Company sold the Kennisis Lake mill to Weldwood of Canada which ran the place in the same manner for a while. Eventually this firm put the prime property around Kennisis and Redstone Lakes for sale for development as cottage lots, and offered the balance of the forest acreage on the open market. The land took several years to tempt a buyer. The assembled package had 60,000 acres, consisting of several buildings, some timber rights, small lakes and a much depleted forest. One overseas businessman took the place for a short while and then sold it to the father of the present owner. When his son Peter Schleifenbaum took over management it was just what this abused property needed, for he held a doctorate in Forestry and soon developed a vision and ongoing plan for revitalizing the Haliburton Forest.

The core of this vision was to build up and restore the depleted resources of the Forest by moving to make its development sustainable for future generations. Roads were cut to make the huge land more accessible and a decision was made to harvest no more than 5% in area and up to 1% of the available timber annually. Tree marking would mean that only trees at optimum for harvest would be taken and horse logging in sensitive areas ensured that the forest floor would be disturbed as little as possible. Simply put these actions would mean that the Forest was treated well. In time the Forest would become multi-use, integrated and sustainable. To achieve this laudable goal, other means of raising revenue had to be built up to finance the whole operation. Whatever happened visitors would see no neon signs. All development would have to be in harmony with the nature of a working forest. The tag 'and Wildlife Reserve' added to the corporate name demonstrated that the stake the original forest inhabitants had in their home would be respected.

Horse logging is enviornmentally and socially beneficial.

Sites were developed for camping in the fifties. Timber management was under way by the present owners in 1972 and snowmobile trails were cut and groomed soon after. A coherent trail system was developed and there are now 300 kilometres of permanently maintained trails. These are complimented for riders by shelter cabins and a handy map of the whole system. One must see excursion is to a frozen waterfall, "the gorge", and another to the lookout at Black Lake, a popular beauty spot. Several special sled events occur during the season. There is a comparable activity for warm weather months developed in the early nineties. Mountain biking, hiking and snowmobiling trails in the Forest have evolved into one of the largest such outdoor adventure opportunities in Eastern North America.

With over 300 kilometers of trails, the Forest is among other things, a snowmobiler's paradise.

Since the late eighties organized outdoor education programs have been developed for both individuals and groups. In a living forest there is a wealth of flora, fauna and wildlife to observe. Orienteering, wilderness survival and naturalist courses are just a few of the opportunities for enjoyment outdoors. Leadership programs, those featuring skills or environmental education attract year round interest. Over the years the Forest has networked with similar local organizations such as the Leslie M. Frost Centre, Medeba Adventure Centre, the Kinark and Kandalore Outdoors Centres and YMCA Wanakita. Education and research partnerships have been forged with Trent University and the University of Toronto, Sir Sandforrd Fleming College and Hocking College of Ohio. To add to this wide range of interests, interns come to the Forest from around the world to learn skills and leadership development.

Amid this good news story disaster struck in the form of a professional forester's worst nightmare in July 1995. An event which the weather people call a tornado down blast struck the revitalized Forest and left a path of devastation in its wake. Five thousand acres were ravaged in a few minutes and the untamed force of nature completely flattened 2,000 of those acres.

There was no choice but to harvest the salvageable timber and wait for the Forest to heal itself. Even today careful observers may note the path of that destructive event.

A tornado down blast flattened 2,000 acres in 1995.

Since the Forest was fast becoming a recognized outdoors mecca, a natural development progression was the construction of the Wolf Centre in 1996. More of this venture later. The next year the Eco-Log concept was announced using hemlock logs. Architect Frank Lloyd Wright might have been referring to this idea when he declared that "Wood is universally beautiful to man. It is the most humanely intimate of all building materials". For 150 years eastern hemlock trees had been used as the core of eastern Ontario barns and other buildings due to its strong nature and resistance to decay and insect attack. Cottage and home building kits of this squared log base have been widely sought for custom designed properties. A mill is used to cut the beams and saw planks to order right in the Forest.

The idea of tree top touring in the Forest became a reality in 1998. The Canopy Tour would become the largest of its kind in the world. Dubbed as a 'Walk in the Clouds', the four hour excursion offers a half kilometre boardwalk tour at up to 75 feet above ground over an old growth pine forest. An ambitious undertaking, the construction with all it's built in safety features took 2,000 worker hours to complete. Each group of 'cloud' travelers take a drive to a lake, a paddle across it and then after instruction in the use of the mandatory safety gear the party slowly ascends through the trees. Part way along the circular pattern walk is a viewing platform enabling a spectacular view of the forest and beyond. One of the reasons this tour is so popular is that it is fully in line with the eco-tourism creed that it educates visitors and yet does not harm the land.

A "Walk in the Clouds".

To add to the activities available in the daylight hours, one may welcome the night at the purpose built observatory. Here not far from the Base Camp complex is the building where the roof slides away on rails. There are three highly sophisticated computer aided Schmidt – Cassegrain telescopes along with an original Star Map by Andrew Hillo which enables star questers' to map their way through the night sky.

A daytime view of the Observatory.

Haliburton Forest has 50 lakes covering in total one tenth of the 60,000 acre reserve. They range in area from 40 to 450 acres. Seven of these waterways are 100 feet or more in depth. In 2004 a small submarine commissioned from a British Columbia company, began taking visitors into one more dimension of nature observation beneath the surface of Macdonald Lake.

There is so much to discover as in pursuits like fishing where the Forest has its own biologists on staff and stocks the lakes without benefit of government intervention. One hundred purebred Siberian Huskies take novice sledders' on cross country runs.

Icy cliffs are the backdrop for a dog sledding adventure.

With at least ten special event weekends in the year and all the other opportunities for personal growth, no wonder some people take a break and retreat to the log cookhouse and admire the art gallery displayed on its walls. In addition to these public pastimes, behind the scenes the Forest is very much involved in scientific assessment and research into its lands and resources. Now widely invested in knowledge, innovation and understanding, this special place has met its original goal of being multi-use, integrated and sustainable.

The Cookhouse restaurant located at the Base Camp.

The Haliburton Forest and Wildlife Reserve has long been recognized by the Forest Stewardship Council of Canada as a truly sustainable forest. This bodes well for wolves. A managed forest protects and plans for forest cover, wetlands, deer yards and trees. All aid the wolf since they are a necessary part of his world.

Learning about good management practices.

The "Adventure Race" is an annual event.

Canoe tripping.

The "Sugar Shack" is an accommodation in the the Base Camp.

A local resident passing through the Base Camp.

Wolves at the Centre

The portion of the Haliburton Forest Wolf Centre used by visitors sits in a small clearing which is also crossed by a sled dog trail. The site chosen was a link with the past as it was formerly used as a log landing, a place where cut wood was gathered and picked up for transport to the mill. The present board and log complex is highlighted by cutouts of three running wolves on the main outside wall of the building. Just by the door a small sign sums up what this place is all about. This is a private educational and research facility funded solely by user fees. Visitors are cautioned to respect the integrity of the facility as well as the privacy of the resident wolf population. Prior to entering and embarking on a unique wildlife experience, it would be wise to reflect on how this place came to be.

The entrance into the Wolf Centre.

In the mid-seventies professional photographer Jim Wuepper set up a wolf research project on Michigan's Upper Peninsula in cooperation with the University of Michigan. This endeavor was carried on for almost two decades. During this period Jim came to know famed naturalist and wildlife author R.D. Lawrence who was producing the documentary 'In Praise of Wolves'. When the project came to an end Jim looked around for a new home for the resident wolf pack. Ron Lawrence made his home in Haliburton and was well aware of the Haliburton Forest mantra of multi-use, integration and sustainability. A wolf facility within the Forest would complement the operation and would also tie in very well with their rapidly developing outdoor educational programs. Ron Lawrence introduced Wuepper to Haliburton Forest and the idea gradually worked out into reality.

Pictured here, Ron Lawrence played an instrumental role in the in the concept of the now renowned Wolf Centre. R.D. Lawrence died in November, 2003.

Paramount in such a venture is the security of the animals. They should not feel threatened by exposure to outside influences and be free to live out their lives much as would other wolves in the Forest. While plans were made to transfer six wolves from one country to another, a 15-acre enclosure was constructed to house them. The habitat inside the enclosure was not altered leaving it a natural forest landscape. This included a small spring-fed pond for access to water which is out of sight to visitors. Along its western perimeter a small observation tower was constructed from which most of the enclosure could be viewed. Saturday afternoon programs were offered to the public which included a weekly feeding. Public interest quickly grew and soon the space inside the tower was no longer sufficient. In 1995 the construction of the present day Wolf Centre on the southern perimeter was complete and a further edition was built in 1998 which houses the yearly themed gallery display. Above all the wolves would be unsocialized, captive and unacclimatized to man and have a continued guarantee that they would be able to live without interaction with humans. Within it they would be free to continue the complex social interaction of the pack.

Since the first six wolves were introduced to Haliburton Forest at Thanksgiving 1993 there has been drama and sadness, hard times and triumphs, birth and death and through the years the pack has renewed and developed just as it would do in the wild. The Centre that fringes the big enclosure has changed and grown to accommodate a continually growing public interest in the life of the pack. After four years of operation it was decided to publish an annual magazine which would chronicle the residents' lives and offer a variety of insights into wolves in general, literature and research on the animals. The annual periodical Wolves has been published ever since and enables visitors to keep in touch with this window into the lives of the members of a wolf pack.

In addition to the observatory there are four other public areas in the Centre. There is a large display area, a combined cinema and classroom, a gallery- which also offers another chance for wolf viewing- and a log building which highlights changing wolf-themed exhibits. The gift shop stocks clothing, books, jewellery and posters, all reflecting aspects of wolves. Those who are frequent visitors to the Centre tend to wander through the displays inbetween viewing the residents from the observatory. They work on the assumption that if the wolves are out of sight for a while, they will be back after a short interval.

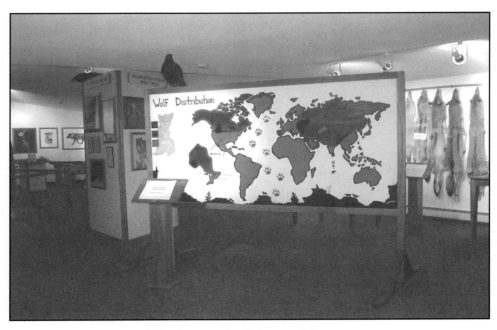

Displays teach visitors wolf lore.

The displays in the Centre add up to one of the most comprehensive collections of diverse materials on wolves in the world. The quest is ongoing to locate more artifacts which further illustrate the wolf and our perception of one of the animal kingdom's least understood species. Depictions of wolves through rare historic prints and statuary are complimented by notions of the animal as seen through the eyes First Nations people's. Visitors are encouraged to make up their own minds about man and his treatment of wolves when faced with furs and the tools of the trapper's trade and then proceed to nearby realistic mounted displays of wolves hunting. Pass through the gallery to the log building and see periodically changed themed exhibitions which have ranged from wolves as toys, being used to sell products through advertising and even the Hollywood view of wolves at the movies. A collection of stamps and special mail covers shows wolf stamps from Hungary, Germany, Czechoslovakia, Poland and the U.S.A.

This Bucharest, Roumania first day stamp cover
demonstrates the interest in European wolves.

Move down to the Observation room and the dominant feature is a wall of one-way viewing glass. A scan outside of the southern exposed hillside reveals trees, rock, and apparently no wolves. The outside directional microphone just picks up wind noise. Look again. Two of the hillocks, bumps on the surface of the uneven area are in reality two nose to tail wolves taking in the early morning sun. Both are asleep but one seems to have an open yellow eye still aware of his surroundings while taking it easy. Patience is a virtue. Wander around and see a statue of St. Francis of Assisi, 1181 to 1226, the patron saint of animals and the man who, as legend has it, made a pact of non violence with a marauding wolf and sealed the deal with a paw shake. Outside there is movement. The two sun worshippers move along to follow the rising sun. Both are a grizzled gray, moving with a lithe, fluid motion. They turn on a dime and settle down to sleep or repose indulging in some form of wolfish reflection.

The observation windows inside the Centre.

Over on a table away from the viewing window are placed albums of pictures and a binder with bulletins of recent events. A blue covered notebook is often missed by visitors and yet within it are staff penciled observations of happenings seen out in the clearing. Along the walls are photographs of current pack members, making identification easy when these players pursue their daily round in front of the unseen audience. Another area of pictures records wolves that have passed on and shows that at least two of the veteran Centre members lived to nine years. Meanwhile maybe those two reclining forms catching some rays in the clearing are dreaming of the large buck enjoyed as the weekly dinner menu a couple of days before. The Alpha animals opened it up and ate first as is their right as leaders. Gradually other wolves joined the feast in order of seniority. In 20 minutes all that remains of the main course of the meal is part of the hide, with a portion of the head and bones left for another time. The main body digested, the cracked open bones will provide both a tasty and nutritious marrow snack.

A glance at the staff notebook reveals a few nuggets of wolf days. Howling has been quite regular with some orchestrations more popular and inclusive than others. One wolf has tried to get a group play going but the others are feeling lazy so the social director quits her efforts. Two yearlings play tug o' war over a well chewed remnant of deer hide. Actions depend on time and mood. Some notes reveal play, others acts of discipline meant to reinforce established notions of pack control. The Alpha male goes after one subordinate who has stepped out of line. He grabs the transgressor's muzzle and the chastised one rolls over, shows his belly in submission and licks the face of the stern figure laying down the law over him. There are more comments about skylarking, plain having fun and every day living than any other behavior.

The sun brightens and the rest of the pack appears and walks around the hillside. There are nine wolves and the watcher counts seven now out in the open. Two exit stage left, presumably to drink at the pond just out of view. Others find a favorite space to sit, relax or groom themselves. Yearlings intently examine weather worn bones and even bat them around. Eventually most of the animals sit and bask in the warm sun. They ignore the glass in front of their accustomed world because they cannot see beyond it. Back in the display area a child has written in the comment book. "What if we called the wolves–and they came?' These wolves did come but they live alone in their own way, not bothering or being touched by humans.

The basement of the Centre is off limits except to staff and those with business there. A lone researcher walks along the hallway and almost bumps into a deer lying on a wheelbarrow. This victim of a highway traffic accident is thawing out and will be the entrée of a wolf dinner soon. Behind him is a walk-in-freezer. Here are more deer and other animals, mostly road kill but some are purchased from trappers interested only in the fur. These carcasses will be food for the wolves at some time in the future. One deer or seven beaver will make up a meal.

Further down the hall the extensive library is a boon to the student of wolves. There are reference books, standard works on the animal, and volumes of natural history. Along with reports of scientific studies are a couple of yards of fantasy, books on werewolves, wolf men and like products of the imagination.

There is a host of children's books, and many shelves of fiction works ranging from themes close to life to pure escapist fun. All relate to the wolf in some way or another. There are videos and magazine articles. The de facto librarian seems to be a small black cat named Matrix who peers through the stacks at the would-be reader. The place is quiet and can be a little unnerving. There is another occupant of the room. From a vantage point in the corner, looking as it were over the reader's shoulder, is a magnificently mounted wolf. The animal was killed on the road but now is posed in next to real life appearance. His expression seems to be saying to the visitor to get those facts right- or else.

A switch on the usual stories that libel wolves.

Largely unnoticed by most visitors is the Centre's role as a background and facilitator for research. This is an ideal place where a wolf pack is held for study and education and beyond the enclosure, the Forest is host to roaming wolf packs. One ongoing study is on the wolf's diet. Roughly two hundred samples of scat, undigested portions of hair, hide and other matter, are collected every year and examined to note just what wolves eat and any changing patterns in their diet. Moose are bedeviled by a parasite called a tick. When wolves eat moose meat the tick's shells may be observed in the scats and give clues toward the prominence of this animal pest. Another study at the Centre is focusing on the behaviour of the unsocialized, captive pack.

Wolves are at home in their element as is demonstrated by Citka.

The pack regularily rests together.

Nishka is a mature female in the Wolf Centre pack.

Citka in the midst of another relaxing yawn as seen from the viewing windows.

Use Your Imagination

Books were scarce in war time England and so a young boy was most pleased to receive a well worn copy of the book Old Peter's Russian Tales. The volume was like the curate's egg, good in parts, but one story stood out among all the rest.

In earlier times some folks were heading for St. Petersburg on a horse-drawn troika. Suddenly a pack of wolves attacked them. One of the men on the back of the big sleigh was not dismayed. He simply reached down inside the mouth of the nearest wolf, pulled him inside out and threw him to his fellows. The wolves stopped to eat this new treat and then this fearless act was repeated several times until the party safely arrived at the city gates.

This fable was one boy's introduction to the world of wolves and many others have similar stories recalling how their interest in wild life was quickened by wolves. Maybe like the boy with the book of fables never thought of a song current at the time. The recurring line went, 'Who's Afraid of the Big Bad Wolf?' Children have ever been some of the first targets of anti-wolf sentiment.

From the ancient times through to relatively modern times, myths and fables have been used to make a point, account for something unknown, instruct the young and even entertain with a nugget of commonsense as a bonus. Cree Indians accounted for human survival after the Flood by having the wolf carry around a ball of moss on which humans might rest until the earth was formed. In the Norse legends the giant Fenris wolf is an offspring of Lokis, the trickster god. Fenris became huge and too hard to control, so the gods had him bound by a magic rope until he broke loose and fought with the gods. Eventually the world was consumed by fire and a new world and race of human beings came into being.

Images of wolves are found everywhere. They are seen on T-shirts, postcards, puzzles, watch faces, blankets and clothing. Outlaw bikers are sometimes called wolves and adopt the emblem to demonstrate their nature. Some men even take the name Wolf as their own. One television program invites watchers to 'Cook with the Wolfman'. No wonder that wolf-related sites on the Internet run into many thousands. Wolves seem to attract the imagination. They thrive in myth and folk tales, literature from epic to lurid, in art, music, sculpture, in sports and the movies. The wolf is accorded the ultimate compliment by being used as a major icon in advertising. Follow along these diverse references to wolves and recall some examples from your own experience.

In a departure from the usual theme, the wolf has designs on Little Red Ridinghood's chocolates.

The Fenris story suggests the danger of trying to restrain earth's wild forces. One more permanent result of the Norse mythology came in words handed down through the ages. Wolfram, the wolf raven, came to be both a name and a hard metallic element, perhaps a tip of the hat to the wolf's resilience and strength. The Norse admired the wolf for his strength and wisdom. Their kings had names like Beowolf (war-wolf), Wulfstan, Wulfred and Ceowulf.

St. George is the patron saint of many factions. In Eastern Europe his day of April 23 honors the wolf. St. Francis, who is celebrated on October 4th, is the patron saint of the animals. In one of his many good works, the saint was approached by villagers who were troubled by the depredations of a fierce wolf. Francis, who seemed to be able to converse with wild creatures at will, made a deal with the wolf. The wolf would not kill livestock and in return the peasants would not hunt him.

February 14th is a day when lovers are commemorated but wolves squeeze in there as well. Among the most famous of the wolf figures is that attributed to the story of the cave of Lupercal. Here Romulus and Remus, the founders of Rome, were suckled and raised by the she-wolf. This led to the Roman feast of Lupercalia and sheep benefited from this celebration as the Horned Fertility god gave them protection from wolf attacks. Festivals like this came together in St. Valentine's Day on February 14th. As for St Valentine, he fell foul of the Roman emperor Claudius II for marrying young men who were thus distracted from their duty to be soldiers and the saint was executed for his marrying ways.

Romulus & Remus of Rome depicted being suckled by a female wolf.

North American Indians have a host of stories involving and associating with wolves and at least fourteen tribes use variations of the word wolf for individual names. The Tewa people of New Mexico tell of wolves and other predators testing early man to make him strong and then sending him into the world. Their version of the resulting human was one dressed in buckskins and carrying a bow and arrows, much like themselves. The Pawnee admired the wolf and tried to emulate him in their hunting. They felt the animal could move like liquid across the plains. He could see 'two looks away' and could hear a cloud passing. When Pawnee hunters put on wolf skins they were calling upon the powers of the wolf. The pelt was used as a cape. Other articles made up a 'wolf bundle' as a representation of the wolf's presence. The Pawnee had a sign for themselves which was the same as that for the wolf. Place the right hand index finger and middle finger in a V next to the right ear, and then bring it forward.

The Indians of the Great Plains admired wolves for their hunting abilities. Here Wolf Robe, a southern Cheyenne Chief wears the Benjamin Harrison peace medal.

Aesop's fables bring the wolf to prominence. One is the familiar story of the boy who cried wolf and as a result folks eventually ignored him. The wolf in sheep's clothing is in similar vein. In another tale the hungry wolf is offered a job by a mastiff guarding children. The wolf wisely looked over the situation and when he saw that the big dog was tied and was occasionally beaten by his masters, decided that his long term freedom was worth more than guaranteed food.

Aesop's story of the Wolf and the Crane relates how a wolf is in pain from a bone stuck in his throat. No animal wanted to help him out for they felt that the bone might have come from one of their relatives or that this was simply a ploy to gain a crane for a meal. Finally the wolf found a helper in a crane when he promised anything if the long necked bird would draw out the bone. She did so and then asked for a reward. The cunning wolf had a surprising answer. She should be grateful he had not bitten off her head when she reached in to get the bone. See a bronze statue depicting this event in the Haliburton Forest Wolf Centre.

In this version of the Aesop fable, the Crane is a scientific type but the supplicant wolf still gets the better of him.

One old story tells of the wolf who had heard that monks have a good life. Those in monasteries lived well, ate lamb and did little work. Against his nature the wily wolf adopted religion and enrolled in the Abbey school. The monk schoolmaster took the new pupil through the alphabet. A and B were soon mastered but when it came to C, the impatient wolf cried out that it must stand for Lamb. Whereupon the wolf was kicked out because he was a humbug.

The Shoshoni people respected the wolf as a creator god who kept the world from death. He could talk and so could coyote, but the people stayed away from the animal they called the Trickster as he was always up to no good. Coyote was jealous of the wolf and tried to put him at odds with the Indian people by saying the earth was overpopulated and that those who died should not be brought back to life. Then one day coyote came seeking the wolf's help because his son was dying of a rattlesnake bite. But the wolf refused, reminding the trickster of his own earlier words. So death came to the people because the wolf never raised anyone from the dead again. The Shoshoni did not hate the wolf and instead admired him for his strength, wisdom and power. As for the coyote, they never have had much time for him.

Among the realm of fairy stories the Three Little Pigs and Little Red Riding Hood stand out. Cautionary tales of the wolf's cunning have remained popular for centuries in varying forms around the world. Lately more politically correct versions have appeared in which the wolf has been framed and even cheated by his erstwhile victims. In long popular tall tales the wolf still leads the field in cunning. A pack of wolves treed a man and could not get him down. A convenient sucker came along in the form of a beaver who was enlisted to cut down the tree refuge. The story goes on from this enterprising beginning.

These make believe pigs are quite happy to sing
"Who's afraid fo the big bad wolf?".

The wolf is most adaptable and has made a leap from fairy stories and fable into the realm of literature. Byron had his pulse on the romantic era. In Mazeppa he writes of The Chase. A pack of wolves is following travelers in the countryside. He paints a word picture of the hunters. They follow with their 'long gallop' and their feet go 'stealing, rustling step repeat'. Saki wrote a fable of nouveau-riche bourgeoisie in which the upstarts are put in their place. They scoff at tales of wolves howling when a member of a noble family dies. But an elderly member of that family, although now on reduced circumstances, states that the wolves never howl for strangers, only those that are truly noble.

Wolves in pursuit of a sleigh was a popular image of the 18th and 19th centuries.

Most fiction writers fall into the trap of anthropomorphism- ascribing human characteristics to animals. Among these is the ability to converse among themselves in English. Rudyard Kipling (1865 – 1936) was one of the most popular writers of his time. The Disney studios have brought his work to millions through several films. Kipling wrote of wolves in The Jungle Book. In the Hunting Song of the Seeonee Pack he pictured them and a howl:

> As the dawn was breaking the Wolf pack yelled
> Once, twice, and again!
> Feet in the jungle that leave no mark!
> Eyes that can see in the dark – in the dark!
> Tongue – give tongue to it! Hark! O hark!
> Once, twice, and again!

Man's cub Mowgli stumbles upon the wolves. Father Wolf carries the child by the scruff of the neck to Mother Wolf. The little one nurses with the wolf cubs and he is adopted by the pack. Farley Mowat's Never Cry Wolf is likely the most well known popular work on wolves in the later half of the twentieth century. He talked with Inuit hunter Ootek about the notion of wolves adopting human children. The consensus was that it was impossible as the child was inherently helpless for too long a time. On the other hand, there were stories among the people of women nursing orphaned wolf pups.

Many scores of books have appeared about wolves in the past century. Overwhelming the sensitive well crafted works by sheer volume are pulp novels among which most authors have never seen a wolf. Jack London by contrast gained millions of readers with his stories of wolves on the frontier. In The Sea Wolf he portrayed Wolf Larsen, a man warring between his brutish and civilized nature. In recent years Clarissa Estes has captured public interest in the animal for her own purposes in Women Who Run With Wolves-Myths and Stories of the Wild Woman Archetype. Stephen King has a different take in The Wolves of Calla, a kind of sci-fi Italo-western.

Sergei Prokofiev (1891-1953) was a most prolific composer but he is mainly remembered today for the hugely popular orchestral fairy tale Peter and the Wolf. As ever, the wolf is portrayed once more as a villain for he ate Peter's duck. The brave lad caught the wolf with a rope and took him to the zoo. One can hear within the music the cries of the duck because the wolf had swallowed him whole and the bird was fortunately still alive.

In a more recent composition, R. Murray Schafer presented his The Princess of the Stars, a musical venture giving credit to the noble nature of the wilderness wolf. When the curious princess hears a wolf howl she falls from the sky and kept prisoner beneath a lake by Three Horned Enemy. Throughout the story the wolf battles to release the princess. The 'music/theatre spectacle' had its Ontario premier at Haliburton Forest at dawn in September 1997, on Wildcat Lake complete with orchestra and a 10 foot high wolf floating on the lake.

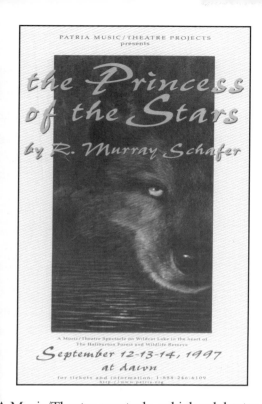 

A Music/Theatre spectacle, which celebrates the character of the wolf, was performed on Wildcat Lake in the heart of the Haliburton Forest and Wildlife Reserve.

Artwork by Curve Lake resident David B. Johnson.

The wolf is no less popular in art. Mongol leader Genghis Khan traced his ancestry back to the grey-blue wolf of heaven. Today the break away Russian Republic of Chechnia features the wolf on its flag and refers to it in its anthem. Nineteenth century painters and lithographers often used wolves in their portraits. One 1894 version of The Wedding Party Attacked by Wolves shows members fighting off the animals with a violin and even their boots. In Toronto's Trinity Bellwoods Park, there is a striking metal sculpture of a wolf howling at the moon. Artistic renderings such as this abound around the world.

Sports teams in soccer, hockey, basketball all use aggressive animals to front their activity. The Sudbury Wolves hockey team and the Minnesota Timber Wolves are two of many. That other entertainment, the movies, has long used the wolf theme. In Wolf, Jack Nicholson plays a middle aged Manhattan editor bitten by a wolf in Vermont who turns into a werewolf. Among the many other popular rendering of wilderness themes, Kevin Costner in Dances With Wolves gained sympathy for the animal and fostered the idea that man might grow from association with them. Company of Wolves, White Fang, The Wolfman, Wolf at the Door, Wolfen, Wolf Call, and even Teen Wolf Too and many others are examples of a variety of portraits of this oft maligned animal.

The movie rekindled an interest in wolves.

Perhaps the ultimate accolade to the wolf and his ability to draw human interest is the way he has been exploited in advertising. "Wolves have ears, don't they?" is a slogan for a radio rock station. Script writers set the wolf up in such diverse poses as an outlaw, an enemy and even the emblem of winter. So it is that the beast that shuns man in the wild is used to sell gas, oil, radiators, perfume, motor bikes, beer, perfume and even sarsaparilla. One airline, Frontier, has a big wolf's head on its tail plane, while limburger cheese has been touted with the slogan 'Keep the wolf from the door'. Some companies have even included wolf in their name, from the inference that the wolf is the leader of the pack.

Follow politicians when seeking references to wolves. A familiar abuse of misguided public perception of wolves is often found in the realm of politics. In the political arena practitioners of this wily art often use allusions to wolves to deride or attack their opponents. They point to real or imagined threats to the electorate or their constituency and embellish them with references to these shy animals that are far removed from human interaction. The enactment of laws relating to wolves has always been subject to the political whim of the moment.

Limburger cheese may have deterred the wolf but it's pungent odour usually has the same effect on humans.

Wolves are commonly used in politics and propaganda, representing the bad and evil as in this 19th century U.S. caricature.

The wolf may justly lay claim to have been exploited through the centuries through many forms of human endeavor. Even as we laud the movement to preserve wolves and hold them up as a symbol of free and unspoiled wilderness, private enterprise uses their identity, naturally without royalty payment.

Canada is often represented as a country with unspoiled wilderness, of which the wolf is a symbol.

This is a depiction of Grimm's tale "The Wolf and The Seven Little Goats".

Few today have heard of playwright Eugene Wolter.

Jack Nicholson's WOLF is seen here in a Japanese playbill.

Tennessee racing driver Jeff Purvis borrows the wolf theme of danger and daring.

Finding Out More

Subscribe to the annual magazine *Wolves* published by Haliburton Forest Wolf Centre, Box 202, RR#1, Haliburton ON Canada K0M 1S0. Look for it to appear around Victoria Day-the third weekend in May.

There are many books about wolves. Here are some of the best adult non fiction.

Barry Holstun Lopez, *Of Wolves and Men*, Scribners, New York, 1978.

Robert H. Busch, *The Wolf Almanac – A Celebration of Wolves and Their World*, Fitzhenry & Whiteside, Toronto.

L. David Mech, *The Way of the Wolf*, Stillwater, MN, 1991.

Books by R.D. Lawrence are especially good for detailed personal observation. Try *Secret Go The Wolves*, Holt, Rinehart, New York, 1980 and *In Praise of Wolves*, Collins, Toronto, 1986.

Ed. D. Landeau, *WOLF – Spirit of the Wild*, Stirling Publishers, The Nature Co., New York.

Candace Savage, *The Nature of Wolves*, Douglas & McIntyre, Vancouver, 1996.

For observation mixed with humor look at possibly the most popular of all well known wolf books, Farley Mowat, *Never Cry Wolf*, McClelland & Stewart, Toronto, 1963.

Those who enjoy tall tales along with their basic information might like James Curran, *Wolves Don't Bite*, Sault Ste. Marie *Star*, 1940. Long out of print this book may be accessed through the Sault Ste. Marie Ontario Public Library.

A book which touches on the Haliburton Forest wolves: Peter Schliefenbaum, Brent Wooton, *The Living Forest*, Haliburton Forest and Wildlife Reserve Press, 2000. Obtain this direct by calling 705-754-2198.

A topical adult fiction book which closely relates to the wolf in the present is Nicholas Evans, *The Loop*, Delacorte, New York, 1998. The author also wrote *The Horse Whisperer*.

Good Juvenile non fiction books include:
D.H.Patent, *Grey Wolf, Red Wolf*, Clarion, New York, 1990.
Anne Welsbacher, *Wolves, Predators of the Wild*, Capstone Press, Mankato, Minnesota USA.

Find some good information on the Internet.
Haliburton Forest & Wildlife Reserve http://www.haliburtonforest.com

The Searching Wolf: this is a listing of interesting wolf and related sites.
Canadian Centre for Wolf Research http://www.wolfca.com

International Wolf Center http://www.wolf.org

Wildlife Science Center http://www.wolftrec.org

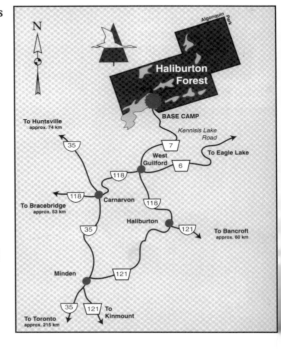

One of the best ways to find out more is to visit Haliburton Forest. There are several ways to come from the Toronto area. One of the most direct is to go north on highway 400 and continue on as it turns into highway 11. At Bracebridge, turn right on highway 118 which is posted to Haliburton. Pass through the crossroads of Carnarvon and come to West Guilford. Turn left, go over the bridge and take county road 7 for 20 kilometres. You can query the Forest by calling 705-754-2198.

Haliburton Forest is just
3 hours north of Toronto.

Support the Wolf

Before thinking about supporting wolves, consider their credo.

>Respect the Elders
>Teach the Young
>Cooperate with the pack
>Play when you can
>Hunt when you must
>Rest in between
>Share your affection
>Voice your feelings
>Leave your mark
>From *WOLVES* issue 2000

Next consider if discussion topics such as these sum up your feelings about wolves.

    Should we continue to snare, trap, and poison and shoot wolves?
    Is the wolf an animal that maintains the balance of nature?
    Has the wolf been made a scapegoat for many human activities?
    Is the wolf a noble animal in need of protection?

If these questions prompt in the reader a need to learn more and even become actively involved in wolf protection and conservation, there are numerous ways to fulfill this interest.

Visit the Haliburton Forest Wolf Centre or any other facility where wolves may be seen free to roam in large enclosures. A wolf howl at Haliburton Forest or a similar event in Algonquin Park will stir the emotions and at the Haliburton Centre the visitor may even see the wolves interact and walk within their forest home. The wolves vocalize in many ways but the howl distinguishes them from all other creatures. Each animal has a distinctive howl which serves to locate, to announce his presence and even send a message. Wolves will respond to human howls and it is great fun.
Watch the media for articles on wolves. The third week in October is

National Wolf Awareness Week and is so celebrated in the United States. This is sponsored by Defenders magazine, an arm of the organization Defenders of Fur Bearers. This group is one of several which have privately donated funds to compensate ranchers and farmers against livestock losses through wolf activity.

Obtain a free copy of your provincial, territorial or state hunting regulations. Use them to become knowledgeable about how wolves may be legally treated and hunted in your area. In 1998 there was no longer a wolf bounty in Ontario but wolves were only protected in Algonquin and Lake Superior Provincial Parks, Nipissing and the Chapleau Crown Game Preserves. They may be killed any time by any one with a small game licence in Southern Ontario and in Northern Ontario between September and June.

John Theberge states that trapping accounts for a minority of deaths. Most are dispatched by neck snares and shooting by people who just do not like wolves. Find out what the position is today on the legal killing of wolves.

Contact organizations that support wolves and promote their well being.
International Wolf Center, 1396 Highway 169, Ely, MN 55731 USA
Canadian Centre for Wolf Research, Box 342,, Shubenacadie, NS B0N 2H0

Many institutions conduct wolf research and seek support for their programs. Contact the Department of Biology at Laurentian University in Sudbury Ontario P3E 2C6 for details of the integrated studies in wildlife program.

As we learn more of wolves, we come to understand our own place within society.

In this 16th century biblical depiction, Jesus is protecting "his flock" from the wolves.

You Were Saying

Here are some well known sayings and facts about wolves. The reader will no doubt want to add many more...

The wolf shall dwell with the lamb.   Isaiah 1:3

Beware of false prophets, which come to you in sheep's clothing, but inwardly they are ravening wolves.   Matthew

Homo homini lupus (Man to man is a wolf)   Plautus

A precipice behind, wolves in front.   Erasmus

The wolf is fed by his feet.   Russian proverb

The wolf doth something every week that keeps him from the church on Sunday.   Old English proverb

Hunger drives the wolf out of the wood.   Fourteenth century French proverb

Give them great meals of beef and iron and steel. They will eat like wolves and fight like Devils.   Shakespeare *Henry V*

Now the hungry lion roars, and the wolf behowls the moon.   Shakespeare *A Midsummer-Night's Dream*

Since all is well, keep it so: wake not a sleeping wolf.   Shakespeare *Henry IV*

He's mad that trusts in the tameness of a wolf, a horse's health, a boy's love, or a whore's oath.   Shakespeare *King Lear*

It disturbs me no more to find men base, unjust or selfish than to see apes mischievous, wolves savage or the vulture ravenous for its prey.   Molliere, *Le Misanthope*

Wolves which batten upon lambs, lambs consumed by wolves, the strong who immolate the weak, the weak victims of the strong: there you have nature, there you have her intentions, there you have her scheme: a perpetual action and reaction, a host of vices, a host of virtues, in one word, a perfect equilibrium resulting from the equality of good and eveil on earth.   Marquis de Sade, *Justime, ou les Malheurs de la Vertu*

It is the caribou that feeds the wolf but it is the wolf that keeps the caribou strong.
Keewatin Innuit saying

Amaguk (the wolf) is like Nunamiut (the people). He doesn't hunt when the weather is bad. He likes to play. He works hard to get food for his family. His hair starts to get white when he is old.   Alaskan Eskimo on parallel between wolf and his people.

The Assyrian came down like the wolf on the fold.
Byron

One learns, said the other, to howl with the wolves.
Racine

The rich are like ravening wolves, who having once tasted human flesh, henceforth desire and devour only men.
Rousseau

The white man must treat the beasts of this land as his brothers. What is man without the beasts? If all the beasts were gone, man would die from a great loneliness of spirit. For whatever happens to the beasts, also happens to the man.
Chief Seattle, Puget Sound Squamish, 1855.

On the ragged edge of the World I'll roam And the home of the Wolf will be my home.
Robert Service

Any man that says he has been et by a wolf is a liar.   Sam Martin, Algoma, 1910

If you live among wolves, you have to act like a wolf.   Nikita Kruschchev, 1971

When anything strengthens a bond of friendship, the friends have walked in the shadow of the rainbow.   Old saying applicable to wolves and man

A wolf is simply a big wild dog, living on flesh that he gets by open chase, recording his call on trees or corner stones, unsuspicious and friendly, wagging his tail for pleasure or baying at the moon.
Ernest Thompson Seaton

There are of course several things in Ontario that are more dangerous than wolves. For instance, the step-ladder.
Jim Curran

If you talk to the animals they will talk with you and you will know each other. If you do not talk to them, you will not know them, and what you do not know you will fear. What one fears one destroys.

                         Chief Dan George

The wolf, now an endangered species, has become a symbol of all that is right and in harmony with nature. It is modern man who in his ignorance has been wrong and out of step with nature. Not the wolf.

                       Michael Fox *The Wolf*

Anonymous has several things to add about wolves:
  Have or hold a wolf by the ears~ to be in a precarious position.
  Throw to the wolves~ sacrifice without compunction.

Expressions which have entered the language include:
  The boy who cried wolf.
  A wolf in sheep's clothing.
  Thrown to the wolves.
  Into the mouth of the wolf. Italian show business equivalent of 'break a leg'.
  Elle a vu le loup..... she has lost her virginity

The habits or nature of wolves has been ascribed to other creatures. A few are wolf eel, wolf moth, wolf spider, wolf fish, wolf snake and wolf wasp.

The Haida Indians call the killer whale the sea wolf.

Wolf's bane or monkshood is a lovely but deadly European plant that got its name originally as a poison for wolves.

The Latin for wolf is the same as that for whore and men who prey on women have been called lone wolves.

Gallows comes from a Saxon word meaning wolf tree.

The name Adolph derives from Edel, the noble wolf. Hitler acknowledged that the wolf captures public imagination. His submarines traveled in wolf packs and one of his retreats was called the Wolf's Lair.

Cancer was often known as the wolf disease. Even today, Lupus in women is incurable.

Acknowledgements

David Bishop and Irene Heaven have both written much about the Haliburton Forest wolves. Irene works as a biologist for the Forest and David co-ordinates outdoor education programs. Peter Schleifenbaum supported this work as part of the ongoing movement of public education about wolves. The staff and interns at Haliburton Forest, especially Paul Brown, have been most helpful and pleasant to the author. Victoria Fraser of Haliburton County Library, Dysart branch, has been as ever resourceful in obtaining books on inter-library loan. Finally thanks to my wife Joan Barnes who once again puts up with, in this case, wolf talk.

Finally, as a plea for all those who live in the wild country:

> What would the world be, once bereft
> Of wet and wilderness? Let them be left.
> O let them be left, wilderness and wet;
> Long live the weeds and the wilderness yet.
> Gerald Manley Hopkins, *Inversnaid*

About the Author

Retired educator, one time newspaper columnist and CBC radio broadcaster, and author of more than 40 books on the north, communities and police work, Michael Barnes, a member of the Order of Canada, lives in the village of Haliburton. His latest seven books are; *The Essential HALIBURTON –Discover Highland Ontario, Dedication to Duty – OPP Officers Who Died Serving Ontario Looking Back KIRKLAND LAKE Haliburton – Memories of Past Years Killer in the Bush – The Great Fires of North Eastern Ontario* and *Looking Back COBALT*.